BEYOND EARTH: THE RISE OF PRIVATE SPACE TRAVEL

By Rayan Bale

Your Gateway to the Stars

THIS BOOK BELONGS TO :

_____

TABLE OF CONTENTS

INTRODUCTION

The cosmos has always fascinated humanity, inspiring dreams of exploring the stars and venturing beyond the confines of our planet. For much of modern history, space exploration has been the domain of government agencies like NASA, ESA, and Roscosmos. These organizations have achieved remarkable feats, from landing on the Moon to sending probes to the far reaches of our solar system. However, a new era is dawning—one that sees private companies leading the charge into the final frontier. Welcome to the world of private space travel, where innovation, ambition, and entrepreneurial spirit are propelling us toward a future once thought to exist only in science fiction.

In recent years, the landscape of space exploration has dramatically shifted. Visionary entrepreneurs like Elon Musk, Jeff Bezos, and Richard Branson have not only dared to dream of reaching the stars but have also laid the groundwork to make these dreams a reality. Companies like SpaceX, Blue Origin, and Virgin Galactic are pioneering new technologies and business models that are revolutionizing how we access and utilize space. Their efforts are not merely expanding our horizons but are also democratizing space travel, making it more accessible and feasible than ever before.

SpaceX, founded by Elon Musk, has been at the forefront of this revolution. With its groundbreaking Falcon 9 rocket and Dragon spacecraft, SpaceX has successfully reduced the cost of launching payloads into orbit. The company's ambitious plans include not only ferrying astronauts to the International Space Station (ISS) but also establishing human colonies on Mars. SpaceX's Starship, a fully reusable spacecraft currently under development, aims to make interplanetary travel a reality, bringing us closer to becoming a multiplanetary species.

Jeff Bezos' Blue Origin is another key player in the private space sector. With its motto "Gradatim Ferociter" (Step by Step, Ferociously), Blue Origin is methodically advancing spaceflight technology. The New Shepard rocket, designed for suborbital flights, has already carried multiple passengers to the edge of space, offering a glimpse of the Earth from above and a taste of the weightlessness of space. Blue Origin's New Glenn, a heavy-lift orbital launch vehicle, promises to further expand the possibilities of commercial space travel, supporting missions ranging from satellite deployment to deep space exploration.

Virgin Galactic, led by Richard Branson, focuses on making space tourism a reality. The company's SpaceShipTwo is designed to take passengers on suborbital journeys, allowing them to experience several minutes of microgravity and view the curvature of the Earth against the backdrop of space. Virgin Galactic's vision extends beyond tourism; it aims to develop high-speed point-to-point travel on Earth, significantly reducing travel times between global destinations.

The rise of private spaceflight is not just about reaching new heights; it is about fostering innovation and collaboration. NASA and other space agencies are increasingly partnering with private companies to achieve their goals. These public-private partnerships are driving down costs, accelerating the development of new technologies, and opening up new opportunities for scientific research and commercial ventures.

As we stand on the brink of this new era, the potential for private space travel is boundless. From the prospect of vacationing in space hotels to the dream of colonizing Mars, the future is brimming with possibilities. This book, "Beyond Earth: The Rise of Private Space Travel," will take you on a journey through the history, technology, economics, and human stories that are shaping the next frontier of exploration. Join us as we explore how private space travel is transforming our world and what it means for the future of humanity. Whether you are a space enthusiast, a budding entrepreneur, or simply curious about what lies beyond our planet, this book will provide you with a comprehensive and engaging look at the revolution happening in our skies.

CHAPTER I:

THE DAWN OF PRIVATE SPACEFLIGHT

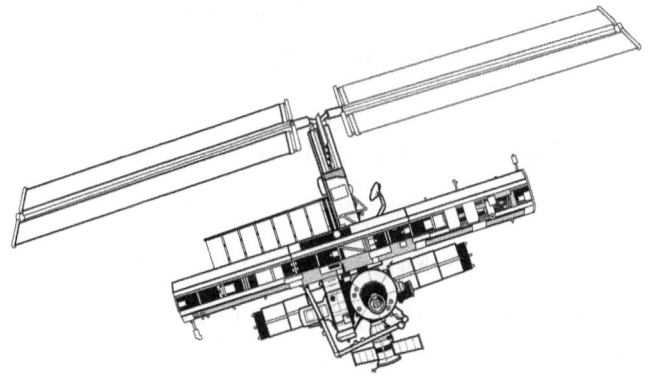

The Dreamers and Innovators:

The quest for space travel has long been a source of inspiration and aspiration for humanity. From the early works of science fiction writers like Jules Verne and H.G. Wells to the thrilling reality of the Apollo missions, space has always represented the ultimate frontier. For many decades, space exploration was the exclusive realm of government agencies. NASA, with its groundbreaking Apollo program, set the bar high with the first human landing on the Moon in 1969. However, the turn of the 21st century witnessed a paradigm shift. Visionary entrepreneurs began to challenge the status quo, driven by a desire to make space accessible to all.

SpaceX: Pioneering the New Space Age:

Elon Musk, the enigmatic founder of SpaceX, is one of the most influential figures in the private space industry. Musk's journey began with a simple yet audacious goal: to reduce the cost of space travel and make Mars colonization possible. Founded in 2002, SpaceX aimed to revolutionize space technology with the ultimate goal of enabling people to live on other planets.

The early years of SpaceX were fraught with challenges. The company's first three launches of the Falcon 1 rocket ended in failure, nearly bankrupting the company. However, Musk's relentless drive and the team's perseverance paid off. On September 28, 2008, the Falcon 1 became the first privately developed liquid-fueled rocket to reach orbit. This success marked a turning point, proving that a private company could achieve what had previously been the domain of government agencies.

SpaceX continued to break new ground with the development of the Falcon 9 rocket, which featured the revolutionary concept of reusability. The ability to land and reuse rockets dramatically reduced launch costs, making space more accessible. In 2015, SpaceX achieved the first successful vertical landing of a Falcon 9 first stage, a milestone that would pave the way for future missions. The Dragon spacecraft, developed by SpaceX, also played a critical role in ferrying cargo to the International Space Station (ISS) and, eventually, transporting astronauts as part of NASA's Commercial Crew Program.

Blue Origin: Gradatim Ferociter:

While SpaceX was making headlines with its rapid advancements, another private space company was quietly making significant strides. Blue Origin, founded by Amazon's Jeff Bezos in 2000, adopted a more methodical approach to space exploration, encapsulated in its Latin motto, "Gradatim Ferociter" (Step by Step, Ferociously).

Blue Origin's first major achievement came with the New Shepard rocket, named after Alan Shepard, the first American in space. Designed for suborbital space tourism, New Shepard is a fully reusable rocket capable of carrying passengers to the edge of space. In 2015, Blue Origin successfully launched and landed the New Shepard booster, marking the first time a rocket had flown to space and returned to Earth intact.

The company's ambitions extend beyond suborbital flights. Blue Origin is developing the New Glenn rocket, a heavy-lift vehicle designed for orbital missions. Named after John Glenn, the first American to orbit Earth, New Glenn aims to compete with SpaceX's Falcon Heavy. With a reusable first stage and the capability to launch large payloads, New Glenn is set to play a significant role in the future of commercial spaceflight.

Virgin Galactic: Making Space Tourism a Reality:

Richard Branson's Virgin Galactic took a different approach to space exploration by focusing on space tourism. Founded in 2004, Virgin Galactic aimed to create a commercial spaceline that would offer suborbital flights to paying customers. The company's SpaceShipTwo, a reusable spaceplane, is designed to carry passengers to the edge of space, where they can experience weightlessness and see the curvature of the Earth.

Despite setbacks, including a tragic accident during a test flight in 2014, Virgin Galactic has persevered. In 2018, SpaceShipTwo, named VSS Unity, reached the edge of space for the first time with two pilots on board. This milestone was a significant step toward making space tourism a reality. Virgin Galactic's approach has generated significant interest and investment, demonstrating that there is a market for commercial space travel experiences.

The Role of NASA and Public-Private Partnerships:

While private companies have made remarkable progress, their success has been bolstered by collaboration with government agencies, particularly

NASA. Recognizing the potential of private spaceflight to complement its missions, NASA has embraced public-private partnerships. Programs like the Commercial Crew Program and the Commercial Resupply Services have leveraged the innovation and efficiency of private companies to achieve mutual goals.

The Commercial Crew Program, for instance, awarded contracts to SpaceX and Boeing to develop spacecraft capable of transporting astronauts to the ISS. SpaceX's Crew Dragon and Boeing's CST-100 Starliner are products of this initiative, designed to provide safe and reliable crew transportation. These partnerships have not only reduced costs for NASA but also accelerated the development of new technologies.

The Future: Mars and Beyond:

As we look to the future, the ambitions of private space companies continue to grow. Elon Musk's vision of a self-sustaining colony on Mars is no longer a distant dream but a concrete goal with ongoing development. SpaceX's Starship, a fully reusable spacecraft designed for interplanetary travel, represents a bold step toward making humans a multiplanetary species. With plans for cargo missions to Mars in the near term and crewed missions to follow, the possibility of human settlement on the Red Planet is inching closer to reality.

Blue Origin's vision also includes lunar exploration and beyond. The company's Blue Moon lander is designed to deliver payloads to the lunar surface, supporting NASA's Artemis program, which aims to return humans to the Moon and establish a sustainable presence. Jeff Bezos envisions a future where millions of people live and work in space, utilizing the vast resources of the solar system to benefit Earth.

Virgin Galactic continues to innovate in the realm of space tourism, with plans to expand its offerings and potentially develop high-speed point-to-point travel on Earth. By making space accessible to more people, Virgin Galactic is helping to foster a broader interest in space exploration and inspire the next generation of space enthusiasts.

Conclusion:

The dawn of private spaceflight marks a transformative period in human history. Visionary entrepreneurs and their companies are pushing the boundaries of what is possible, making space travel more accessible, affordable, and sustainable. Through innovation, collaboration, and an unyielding drive to explore, private space companies are not only achieving remarkable milestones but also inspiring a new generation to look to the stars.

As we embark on this journey "Beyond Earth," the possibilities are boundless. The rise of private space travel is opening new frontiers, not just for exploration but for commerce, science, and human experience. This chapter has only just begun, and the stories that follow will undoubtedly be filled with adventure, discovery, and the realization of dreams once thought impossible. Welcome to the future of space travel.

CHAPTER 2:

MAJOR MILESTONES
AND MISSIONS

Early Milestones and Achievements:

The journey of private space companies has been marked by significant milestones that have reshaped the landscape of space exploration. SpaceX, founded by Elon Musk, made history with its first successful orbital launch of the Falcon 1 rocket on September 28, 2008. This achievement proved that a privately funded company could achieve the same feats as government space agencies. The success of the Falcon 1 paved the way for the development of more advanced rockets, such as the Falcon 9.

Blue Origin, founded by Jeff Bezos, achieved a groundbreaking milestone in 2015 with the New Shepard rocket. This was the first rocket to successfully achieve vertical landing and reusability after a suborbital flight. The ability to reuse rockets significantly reduces the cost of space travel and represents a major technological advancement in the industry.

Virgin Galactic, under the leadership of Richard Branson, reached a significant milestone in December 2018 when its SpaceShipTwo, VSS Unity, completed its first successful flight to the edge of space with two pilots on board. This achievement demonstrated the viability of commercial space tourism and generated significant public interest in the potential for private space travel.

ISS Resupply and Crewed Missions:

SpaceX has played a crucial role in supporting the International Space Station (ISS) through NASA's Commercial Resupply Services (CRS) program. Since 2012, SpaceX's Dragon spacecraft has been delivering essential supplies and equipment to the ISS. The Dragon spacecraft's ability to return to Earth with scientific samples and equipment has been invaluable for ongoing research and operations aboard the ISS.

The evolution of the Dragon spacecraft into the Crew Dragon variant marked a new era in human spaceflight. In May 2020, the Crew Dragon spacecraft successfully transported NASA astronauts Robert Behnken and Douglas Hurley to the ISS, marking the first crewed mission launched by a private company. This historic mission, known as Demo-2, validated the capabilities of SpaceX's Crew Dragon and demonstrated the potential for private companies to play a key role in human space exploration.

Blue Origin has also contributed to the ISS through its development of the Blue Moon lander, which is designed to support lunar missions. The Blue Moon lander aims to deliver large payloads to the lunar surface, facilitating NASA's Artemis program, which seeks to return humans to the Moon and establish a sustainable presence there. Blue Origin's efforts in developing lunar infrastructure are crucial for future missions to the Moon and beyond.

Lunar and Mars Missions:

SpaceX's ambitious plans for Mars colonization are centered around the development of the Starship spacecraft. Starship is designed to be a fully reusable spacecraft capable of carrying large numbers of passengers and cargo to destinations such as the Moon, Mars, and beyond. The iterative development process of Starship has included numerous test flights, each providing valuable data and insights for improving the spacecraft's design and capabilities. SpaceX aims to conduct regular missions to Mars, with the ultimate goal of establishing a self-sustaining human colony on the Red Planet.

Blue Origin's vision for lunar exploration includes the Blue Moon lander, which is designed to deliver significant payloads to the lunar surface. The Blue Moon lander will support NASA's Artemis program and contribute to the establishment of a sustainable human presence on the Moon. Blue Origin's focus on developing lunar infrastructure and in-situ resource utilization (ISRU) technologies will be critical for future lunar missions and the broader goal of human expansion into the solar system.

Virgin Galactic's focus on space tourism does not preclude its involvement in scientific and exploratory missions. The company plans to expand its offerings to include research missions and high-speed point-to-point travel on Earth. Virgin Galactic's innovative approach to suborbital flights and space tourism will continue to generate public interest and investment in space exploration.

Public-Private Partnerships: A New Model for Space Exploration:

The collaboration between NASA and private companies has proven to be a successful model for advancing space exploration. NASA's Commercial Crew Program has been instrumental in developing safe and reliable crew transportation systems. By awarding contracts to SpaceX and Boeing, NASA has fostered innovation and competition in the private space sector. SpaceX's Crew Dragon and Boeing's CST-100 Starliner are products of this initiative, designed to provide safe and reliable crew transportation to the ISS.

These public-private partnerships have not only reduced costs for NASA but also accelerated the development of new technologies. The success of these collaborations highlights the potential for continued cooperation between public and private entities in space exploration. The partnership model leverages the strengths of both sectors, combining the innovation and efficiency of private companies with the experience and resources of government agencies.

Future Plans and Potential Missions

The future of private spaceflight is filled with ambitious plans and potential missions. SpaceX aims to conduct regular missions to Mars, with the long-term goal of establishing a self-sustaining colony. The development of the Starship spacecraft is central to this vision, with plans for cargo missions to Mars in the near term and crewed missions to follow.

Blue Origin is focused on building infrastructure to support human presence on the Moon and beyond. The company's Blue Moon lander will play a critical role in delivering payloads to the lunar surface, supporting NASA's Artemis program and future lunar missions. Blue Origin's long-term vision includes developing technologies for in-situ resource utilization, which will be essential for sustaining human presence on the Moon and other celestial bodies.

Virgin Galactic continues to innovate in the realm of space tourism, with plans to expand its offerings and potentially develop high-speed point-to-point travel on Earth. The company's SpaceShipTwo and future spacecraft will provide opportunities for scientific research, commercial ventures, and public engagement in space exploration. Virgin Galactic's innovative approach to suborbital flights will continue to inspire and attract interest in space tourism and exploration.

Conclusion:

The milestones and missions achieved by SpaceX, Blue Origin, and Virgin Galactic have transformed the landscape of space exploration. These companies have demonstrated that private enterprises can achieve remarkable feats, from resupplying the ISS to planning missions to Mars and the Moon. As they continue to innovate and collaborate with government agencies, the future of space exploration looks promising and filled with potential. The next decade will likely see even more groundbreaking achievements, further solidifying the role of private companies in the quest to explore the cosmos.

CHAPTER 3:

TECHNOLOGICAL

INNOVATIONS

Rocket Technology Advancements:

Reusability: The Revolution in Rocket Design:

One of the most groundbreaking advancements in the field of rocketry is the development of reusable rockets. Traditionally, rockets were single-use, with each launch resulting in the loss of expensive hardware. This model was not only costly but also unsustainable for the long-term goals of space exploration. SpaceX revolutionized this paradigm with the introduction of the Falcon 9 rocket, which features a reusable first stage. The first successful landing and recovery of a Falcon 9 first stage occurred on December 21, 2015, marking a significant milestone in space technology.

The reusability of rockets drastically reduces the cost per launch, making space access more affordable and frequent. SpaceX's iterative design and testing process for reusability has set a new standard in the industry. The ability to land and reuse rockets has proven to be a game-changer, influencing other companies to adopt similar technologies. Blue Origin's New Shepard rocket, designed for suborbital missions, also achieves vertical landing and reusability, further demonstrating the viability of this technology.

Advanced Propulsion Systems:

Advancements in propulsion systems have been pivotal in enhancing the efficiency and capabilities of modern rockets. SpaceX has achieved significant advances in rocket engine tech for Starship, marking a leap forward. These engines use methane and oxygen, improving efficiency and sustainability. Methane production on Mars aligns with SpaceX's goals for colonization, making these advancements crucial.

Blue Origin's propulsion tech for New Glenn uses liquid oxygen and LNG. These engines are reusable and highly efficient, providing strong thrust for various missions. This makes them suitable for both orbital and suborbital flights, enhancing the rocket's capabilities. These advancements in propulsion systems are crucial for the next generation of space missions, enabling longer durations and higher payload capacities.

Material Science and Manufacturing Techniques:

The use of new materials and manufacturing methods has critically advanced rocket tech. Lightweight, strong materials like carbon composites and advanced alloys are now common in rocket construction. These provide excellent performance and durability, vital for the harsh conditions of space travel.

3D printing, or additive manufacturing, has revolutionized rocket component production. SpaceX and other firms use it to create complex engine parts, cutting manufacturing time and costs while boosting precision and reliability. This tech supports rapid prototyping and iteration, speeding up development and enabling more efficient and advanced rocket designs.

Development of New Spacecraft:

SpaceX's Dragon and Starship:

SpaceX has made significant strides in spacecraft design with the development of the Dragon and Starship spacecraft. The Dragon spacecraft, initially designed for cargo missions to the ISS, has evolved into the Crew Dragon variant, capable of carrying astronauts. Crew Dragon features modern avionics, advanced life support systems, and an innovative launch escape system that enhances crew safety during ascent.

Starship, currently under development, represents SpaceX's vision for a fully reusable spacecraft capable of carrying large numbers of passengers and cargo to destinations such as the Moon, Mars, and beyond. The spacecraft is designed to perform a variety of missions, from satellite deployment to interplanetary exploration. Starship's ability to refuel in space is a key feature, enabling longer missions and the potential for sustainable human presence on other planets.

Blue Origin's New Shepard and New Glenn:

Blue Origin's New Shepard rocket and capsule are designed for suborbital missions, primarily targeting the space tourism market. The New Shepard system consists of a reusable booster and a crew capsule that can carry passengers to the edge of space, providing a few minutes of weightlessness and breathtaking views of Earth. The successful development and operation of New Shepard have demonstrated the feasibility of commercial suborbital flights.

The New Glenn rocket, named after astronaut John Glenn, is Blue Origin's heavy-lift orbital launch vehicle. Designed to be reusable, New Glenn is intended for a wide range of missions, including satellite deployment and interplanetary missions.

The rocket's large payload capacity and advanced design make it a competitive option in the commercial launch market.

Innovations in Life Support Systems:

Developing reliable life support systems is critical for long-duration space missions. Advances in this area include closed-loop systems that recycle air, water, and waste, reducing the need for resupply missions and enhancing sustainability. SpaceX's Crew Dragon and Blue Origin's planned lunar habitats incorporate advanced life support technologies to ensure the safety and well-being of astronauts during missions.

NASA's Artemis program, supported by companies like Blue Origin, also focuses on developing sustainable life support systems for lunar habitats. These systems will enable longer stays on the Moon and serve as a testing ground for technologies needed for future Mars missions. Innovations in life support systems are essential for ensuring that astronauts can live and work in space for extended periods.

Space Habitats and In-Space Manufacturing:

Space Habitats: Building a Home in Space

As humanity ventures further into space, the development of space habitats becomes increasingly important. These habitats provide living and working environments for astronauts on long-duration missions. Concepts for space habitats range from inflatable modules, like those developed by Bigelow Aerospace, to rigid structures designed for deep-space missions.

NASA's Lunar Gateway project, a planned space station orbiting the Moon, will serve as a habitat and staging point for lunar missions. The Gateway will provide living quarters, laboratories, and docking ports for spacecraft, supporting long-term exploration of the lunar surface. Private companies are also exploring concepts for space habitats that can support commercial activities, research, and tourism.

In-Space Manufacturing: Building Beyond Earth:

In-space manufacturing is an emerging field that has the potential to revolutionize space exploration. The ability to manufacture parts and equipment in space reduces the need to transport everything from Earth, lowering costs and increasing mission flexibility. Additive manufacturing is a key technology in this area, enabling the production of complex components in microgravity.

NASA and private companies are conducting experiments on the ISS to advance in-space manufacturing techniques. These experiments focus on producing everything from tools to habitats, demonstrating the potential for on-demand manufacturing in space. The development of in-space manufacturing capabilities will be crucial for future missions to Mars and beyond, where resupply from Earth is limited.

The Future of Space Technology:

Next-Generation Propulsion Systems:

Looking to the future, next-generation propulsion systems are being developed to enable faster and more efficient space travel. Concepts such as nuclear thermal propulsion (NTP) and ion propulsion offer higher efficiencies and longer mission durations compared to conventional chemical rockets. These advanced propulsion systems are critical for missions to distant destinations, such as Mars and the outer planets.

NASA and private companies are investing in research and development of these propulsion technologies. NTP, for example, uses nuclear reactions to heat a propellant, providing significantly higher thrust and efficiency than chemical rockets. Ion propulsion, on the other hand, uses electric fields to accelerate ions, offering efficient and sustained thrust for long-duration missions.

Artificial Intelligence and Automation:

Artificial intelligence (AI) and automation are playing an increasingly important role in space exploration. AI systems are used to analyze vast amounts of data collected by spacecraft, identify patterns, and make real-time decisions. Automation enables spacecraft to perform complex tasks autonomously, reducing the need for constant human intervention.

AI and automation are particularly valuable for missions to distant destinations, where communication delays make real-time control difficult. These technologies are being integrated into spacecraft for navigation, landing, and scientific research, enhancing the capabilities and safety of missions.

Interplanetary Internet and Communication:

Reliable communication is essential for space missions, and advancements in this area are critical for future exploration. The concept of an interplanetary internet involves creating a network of satellites and communication relays that enable continuous data transmission between Earth, spacecraft, and other planets. This network would ensure that missions to Mars and beyond have reliable and high-speed communication links.

NASA's Deep Space Network (DSN) is a key component of current space communication infrastructure, providing communication support for missions beyond Earth's orbit. Future developments aim to expand this network and integrate new technologies, such as laser communication, which offers higher data transmission rates compared to traditional radio waves.

Conclusion:

The technological innovations in rocket design, propulsion systems, spacecraft development, life support, and in-space manufacturing are revolutionizing space exploration. Companies like

SpaceX, Blue Origin, and Virgin Galactic are at the forefront of these advancements, driving the industry forward and making space more accessible. As we continue to push the boundaries of what is possible, these innovations will pave the way for future exploration, settlement, and commercial activity in space.

The future of space technology is bright, with next-generation propulsion systems, AI and automation, and advanced communication networks set to transform our capabilities. These advancements will not only enable us to explore new frontiers but also ensure that humanity can thrive beyond Earth. The journey to the stars is just beginning, and the innovations of today will shape the future of space exploration for generations to come.

CHAPTER 4:

THE ECONOMICS OF

SPACE TRAVEL

Funding and Investment in the Private Space Sector:

Venture Capital and Private Investment:

The growth of the private space sector has been fueled by substantial investments from venture capital firms and private investors. Over the past decade, billions of dollars have poured into space startups, driving innovation and expansion. Companies like SpaceX, Blue Origin, and Virgin Galactic have benefited from significant funding rounds, allowing them to pursue ambitious projects and develop advanced technologies. Venture capital has played a crucial role in transforming the space industry from a government-dominated domain to a vibrant, competitive marketplace.

SpaceX, for example, has raised over $5 billion through various funding rounds. These investments have enabled the company to develop and launch reusable rockets, drastically reducing the cost of space access. Similarly, Blue Origin, funded primarily by Jeff Bezos, has received substantial investments to develop reusable rockets and lunar landers. Virgin Galactic went public in 2019 through a merger with a special-purpose acquisition company (SPAC), raising significant capital to fund its space tourism ventures.

Government Contracts and Grants:

In addition to private investment, government contracts and grants have been instrumental in supporting the private space sector. NASA's Commercial Crew and Cargo Programs have awarded multibillion-dollar contracts to companies like SpaceX and Boeing to develop and operate spacecraft for missions to the International Space Station (ISS). These contracts provide a reliable revenue stream and help mitigate the financial risks associated with developing new technologies.

The U.S. government has also provided grants and incentives to support space exploration and innovation. The Small Business Innovation Research (SBIR) program, for example, offers funding to small businesses developing innovative space technologies. These government initiatives not only provide financial support but also foster collaboration between public and private entities, driving advancements in space exploration.

Public Offerings and Crowdfunding:

Public offerings and crowdfunding have emerged as alternative funding sources for space companies. Virgin Galactic's public listing in 2019 through a SPAC merger provided the company with access to public markets and additional capital for its space tourism operations. This move marked a significant milestone, demonstrating that space companies could attract substantial investments from public markets.

Crowdfunding platforms have also been used to raise funds for specific space projects. For example, Planetary Resources, a company focused on asteroid mining, launched a crowdfunding campaign to support its Arkyd-100 telescope. These platforms enable smaller investors to participate in the space sector, democratizing access to investment opportunities and raising public awareness about space exploration.

Economic Impact of Space Tourism and Commercial Spaceflight:

Space Tourism: A New Frontier:

Space tourism represents a significant economic opportunity within the private space sector. Companies like Virgin Galactic, Blue Origin, and SpaceX are pioneering efforts to make space travel accessible to private individuals. Virgin Galactic, with its SpaceShipTwo spacecraft, offers suborbital flights that provide passengers with a few minutes of weightlessness and stunning views of Earth. Blue Origin's New Shepard rocket provides a similar experience, targeting affluent individuals seeking a unique adventure.

The potential market for space tourism is substantial. Analysts project that the space tourism industry could generate billions of dollars in revenue over the next decade. High-profile flights, such as those undertaken by Richard Branson and Jeff Bezos in 2021, have garnered significant media attention, highlighting the viability and appeal of commercial space travel.

Commercial Spaceflight and Satellite Deployment:

Beyond space tourism, commercial spaceflight and satellite deployment constitute major revenue streams for private space companies. The demand for satellite launches has increased with the growth of telecommunications, Earth observation, and global positioning systems (GPS). SpaceX's Falcon 9 rocket, known for its cost-effectiveness and reliability, has become a preferred choice for launching satellites into orbit. The company's Starlink project, aimed at creating a global satellite internet network, exemplifies the commercial potential of satellite deployment.

Blue Origin and other companies are also entering the satellite launch market, developing rockets capable of deploying payloads into various orbits. This competition drives innovation and reduces costs, making space more accessible for commercial ventures. The increasing number of satellite launches contributes to the growth of the global space economy, estimated to reach $1 trillion by 2040.

Economic Benefits to Society:

The economic benefits of space exploration extend beyond the revenues generated by private companies. Advances in space technology have led to numerous innovations with applications on Earth. For instance, satellite technology has revolutionized telecommunications, weather forecasting, and disaster management. GPS, originally developed for military use, has become an integral part of daily life, enabling navigation and location-based services.

Moreover, the development of new materials and manufacturing techniques for space exploration has spurred innovation in other industries. For example, 3D printing technology, initially used for manufacturing rocket components, is now widely adopted in various sectors, including healthcare and construction. The economic impact of these technological advancements underscores the value of investing in space exploration.

Cost Reduction Strategies:

Reusable Rockets and Spacecraft:

One of the most effective strategies for reducing the cost of space travel is the development of reusable rockets and spacecraft. SpaceX's Falcon 9 rocket and Crew Dragon spacecraft are designed to be reused multiple times, significantly lowering the cost per launch. The ability to recover and refurbish rockets reduces the need for new manufacturing, making space missions more economically viable.

Blue Origin's New Shepard rocket also exemplifies the benefits of reusability. By recovering and reusing the booster and crew capsule, Blue Origin can offer suborbital flights at a lower cost. The continued development of reusable technologies will be crucial for making space travel affordable and sustainable.

Efficient Manufacturing and Supply Chains:

Improving manufacturing processes and optimizing supply chains are essential for reducing costs in the space industry. Companies like SpaceX have embraced vertical integration, producing many of their components in-house to maintain quality control and reduce expenses. Additive manufacturing, or 3D printing, is another cost-saving technique that allows for the rapid production of complex parts with minimal waste.

Supply chain optimization also plays a critical role in cost reduction. By establishing efficient logistics and inventory management systems, space companies can minimize delays and reduce the costs associated with sourcing materials and components. Collaboration with suppliers and the adoption of lean manufacturing principles further enhance efficiency and cost-effectiveness.

Economies of Scale

As the demand for space missions increases, companies can achieve economies of scale, reducing the cost per unit of production. SpaceX's ambitious launch schedule for its Starlink satellites, for example, allows the company to spread the fixed costs of rocket development and manufacturing across multiple missions. This approach lowers the overall cost of each launch and enables SpaceX to offer competitive pricing for satellite deployment.

The expansion of the commercial space market, driven by space tourism, satellite deployment, and scientific missions, will further enhance economies of scale. As more missions are conducted and more rockets and spacecraft are produced, the cost per mission will continue to decrease, making space access more affordable for a broader range of customers.

Conclusion:

The economics of space travel are evolving rapidly, driven by substantial investments, innovative cost-reduction strategies, and the expanding market for commercial spaceflight and space tourism. Private companies like SpaceX, Blue Origin, and Virgin Galactic are at the forefront of this transformation, leveraging venture capital, government contracts, and public offerings to fund their ambitious projects.

The economic impact of space exploration extends beyond the revenues generated by these companies, contributing to technological advancements and societal benefits. As the industry continues to grow, the cost of space travel will decrease, making space more accessible and opening new opportunities for exploration, research, and commercial ventures.

The future of the space economy looks promising, with the potential to revolutionize how we interact with space and benefit from its vast resources. The innovations and investments made today will shape the future of space exploration, driving economic growth and inspiring the next generation of space pioneers.

CHAPTER 5:

HUMAN SPACEFLIGHT AND TOURISM

Experiences of Private Astronauts and Space Tourists:

Private spaceflight has opened the doors to an era where not only trained astronauts but also private individuals can journey to the stars. These pioneers of private space travel have had transformative experiences that have captured the public's imagination and provided valuable insights into the potential for commercial space travel.

First-Hand Accounts:

The experiences of private astronauts and space tourists have been nothing short of extraordinary. Space tourists, such as those flying with companies like SpaceX and Blue Origin, have described the sheer awe of viewing Earth from space and the unique sensation of weightlessness. These journeys are not only personal milestones but also significant achievements for the companies facilitating them.

For instance, when SpaceX launched its first all-civilian mission, Inspiration4, the crew members shared their profound experiences and the mission's impact on their lives. They highlighted the beauty of Earth from orbit and the importance of space travel for scientific and philanthropic purposes. Similarly,

Blue Origin's New Shepard flights have provided suborbital experiences that combine adventure with educational outreach, inspiring participants to share their stories widely.

Scientific and Personal Achievements:

Private space missions are not just for thrill-seekers; they also contribute to scientific research. Private astronauts often participate in experiments designed to advance our understanding of space's effects on the human body, materials, and technology. These missions bridge the gap between adventure and meaningful scientific contributions.

The success stories of these missions are not only about reaching space but also about achieving personal growth and contributing to a larger narrative of human space exploration. Private astronauts often return with a renewed sense of purpose and a deeper appreciation for the planet and the potential of space travel.

Training and Preparation for Space Travel:

Preparing for space travel involves rigorous training to ensure the safety and well-being of private astronauts. This preparation is crucial for both suborbital and orbital missions, providing participants with the knowledge and skills needed to handle the challenges of spaceflight.

Training Programs:

Training programs for private astronauts are comprehensive, covering various aspects of space travel. These programs typically include physical conditioning, simulations of spaceflight conditions, and training in the use of spacecraft systems. Participants undergo medical evaluations to ensure they are fit for the physical demands of space travel.

For example, SpaceX's training for the Inspiration4 crew included centrifuge training to simulate G-forces, mission simulations, and emergency preparedness drills. Blue Origin's training for New Shepard passengers includes instruction on safety procedures, communication protocols, and experiencing weightlessness through parabolic flights.

Physical and Mental Preparation:

Space travel is demanding both physically and mentally. Private astronauts must be prepared for the intense G-forces experienced during launch and re-entry, the microgravity environment of space, and the confined conditions of spacecraft. Training programs address these challenges through a combination of physical fitness routines and psychological preparation.

Mental resilience is equally important. Space travelers must be able to handle the isolation and stress associated with space missions. Training often includes team-building exercises and scenarios designed to prepare participants for the psychological aspects of space travel.

Impact of Space Tourism on Public Interest and Industry Growth:

The rise of space tourism has had a profound impact on public interest in space exploration and the growth of the space industry. The excitement generated by private spaceflights has captured the imagination of people around the world and inspired a new generation of space enthusiasts.

Public Imagination and Inspiration:

Space tourism has reignited public fascination with space exploration. High-profile missions, such as those undertaken by SpaceX, Blue Origin, and Virgin

Galactic, have received extensive media coverage, bringing space travel into the public spotlight. These missions demonstrate that space is not just the domain of professional astronauts but is becoming accessible to a broader audience.

This renewed interest in space has positive effects on education and career choices, encouraging young people to pursue studies and careers in STEM (science, technology, engineering, and mathematics) fields. Programs that highlight the achievements of private astronauts and space tourists serve to inspire future generations of scientists, engineers, and explorers.

Economic and Social Impacts:

The growth of the space tourism industry has significant economic implications. The demand for space tourism services drives investment in space technology and infrastructure, leading to advancements that benefit the entire space industry. The influx of private capital into space companies supports the development of new technologies and lowers the cost of space access.

Additionally, space tourism promotes international collaboration and cultural exchange. As individuals from different countries participate in space missions, they foster a sense of global unity and shared interest in exploring the cosmos. This cultural exchange enriches the experience of space travel and underscores the universal appeal of space exploration.

Conclusion:

Chapter 5 delves into the transformative experiences of private astronauts and space tourists, the rigorous training and preparation required for space travel, and the profound impact of space tourism on public interest and industry growth. These aspects collectively highlight the importance of human spaceflight and tourism in shaping the future of space exploration. As private space travel continues to evolve, it will undoubtedly inspire new generations and contribute to the ever-expanding frontier of human achievement in space.

CHAPTER 6:

CHALLENGES AND
CONTROVERSIES

Safety Concerns and Regulatory Challenges:

As the private space industry continues to grow, safety concerns and regulatory challenges remain at the forefront. Ensuring the safety of astronauts, space tourists, and the general public is paramount for the continued success and acceptance of commercial space travel.

Safety Standards in Space Travel:

The establishment of rigorous safety standards is crucial to protect human life during space missions. Companies like SpaceX, Blue Origin, and Virgin Galactic invest heavily in developing and maintaining these standards. From the initial design phases to post-mission evaluations, every aspect of a mission is scrutinized to ensure maximum safety.

For instance, SpaceX's Crew Dragon spacecraft underwent extensive testing to validate its safety systems, including launch abort capabilities designed to protect astronauts in the event of an emergency. Similarly, Blue Origin's New Shepard rocket incorporates numerous safety features, such as an escape system to quickly distance the crew capsule from any potential danger during launch.

Regulatory Frameworks and Compliance:

The regulation of space travel is complex, involving multiple agencies and international cooperation. In the United States, the Federal Aviation Administration (FAA) is responsible for licensing commercial space launches and ensuring compliance with safety regulations. The FAA's Office of Commercial Space Transportation oversees the industry, working to balance the promotion of space commerce with the need to protect public safety.

Internationally, space law is governed by treaties such as the Outer Space Treaty of 1967, which outlines the principles for the peaceful use of outer space. Compliance with these treaties and collaboration with international bodies, such as the United Nations Committee on the Peaceful Uses of Outer Space (COPUOS), are essential for maintaining a cohesive regulatory environment.

Addressing Safety Incidents and Risks:

Despite rigorous safety measures, space travel inherently involves risks. The private space industry must be prepared to address safety incidents promptly and transparently. Investigating and learning from accidents is crucial to improving safety protocols and preventing future occurrences.

For example, the tragic accident involving Virgin Galactic's SpaceShipTwo in 2014 highlighted the dangers associated with space tourism. The incident prompted a thorough investigation by the National Transportation Safety Board (NTSB), leading to enhanced safety measures and design improvements. Continuous monitoring and updating of safety procedures are essential to maintaining public trust and ensuring the long-term viability of commercial space travel.

Ethical Considerations in Commercial Space Travel:

The commercialization of space travel raises several ethical questions that need to be addressed to ensure the responsible and equitable use of space resources.

Exploitation of Space Resources:

As private companies explore the potential of mining asteroids and other celestial bodies for resources, ethical concerns about the exploitation of space arise. The commercialization of space resources must be balanced with the need to preserve the space environment and avoid over-exploitation. Establishing ethical guidelines and international agreements on the responsible use of space resources is crucial to prevent conflicts and ensure sustainable practices.

Equity and Access to Space:

Space travel has traditionally been the domain of well-funded government agencies and wealthy individuals. As the industry grows, it is important to consider how to make space more accessible to a broader range of people. Initiatives aimed at democratizing space travel, such as reducing costs and increasing opportunities for education and training, are essential to ensure that the benefits of space exploration are shared more equitably.

Programs like SpaceX's Inspiration4, which included a diverse crew of private astronauts, demonstrate a commitment to making space accessible to individuals from different backgrounds. Such initiatives help to inspire future generations and promote the idea that space is a shared frontier for all humanity.

Ethical Implications of Space Colonization:

The potential colonization of other planets, particularly Mars, raises ethical questions about human impact on extraterrestrial environments. As companies like SpaceX pursue plans for Mars colonization, it is important to consider the implications of human presence on these environments. Preserving the integrity of extraterrestrial ecosystems and ensuring that human activities do not cause irreversible harm are critical ethical considerations.

Additionally, the governance of space colonies presents ethical challenges. Developing fair and just systems of governance for space colonies, including laws that protect the rights of settlers and promote peaceful coexistence, is essential to avoid conflicts and ensure the responsible expansion of human presence in space.

Space Debris and Environmental Impact:

The increase in space activities has led to growing concerns about space debris and the environmental impact of rocket launches. Addressing these issues is critical to ensuring the long-term sustainability of space exploration.

The Problem of Space Debris:

Space debris, also known as space junk, consists of defunct satellites, spent rocket stages, and other fragments resulting from collisions and disintegration of spacecraft. This debris poses a significant risk to operational satellites, the International Space Station, and future space missions. The proliferation of space debris increases the likelihood of collisions, which can create even more debris in a dangerous cascade effect known as the Kessler syndrome.

Efforts to mitigate space debris include developing technologies for active debris removal and designing satellites with deorbiting mechanisms to ensure they re-enter the Earth's atmosphere safely at the end of their operational lives. International cooperation and adherence to guidelines set by organizations like the Inter-Agency Space Debris Coordination Committee (IADC) are essential to addressing this global challenge.

Environmental Impact of Rocket Launches:

Rocket launches have environmental consequences, including the release of greenhouse gases and other pollutants into the atmosphere. The use of certain rocket propellants can contribute to ozone layer depletion and other environmental issues. As the frequency of launches increases, it is important to develop and adopt more environmentally friendly propulsion technologies.

Companies like SpaceX and Blue Origin are exploring the use of cleaner fuels and developing reusable launch systems to minimize environmental impact. For example, SpaceX's Raptor engines use methane, which burns cleaner than traditional rocket fuels and can potentially be produced sustainably on Mars.

Sustainable Practices and Innovations to Mitigate Impact:

Sustainability in space exploration involves adopting practices that minimize the environmental impact of space activities. This includes developing reusable rockets, using greener propellants, and implementing debris mitigation strategies. Innovative approaches, such as in-space manufacturing and resource utilization, also contribute to sustainability by reducing the need for resource-intensive launches from Earth.

Promoting sustainability in space exploration requires collaboration between governments, private companies, and international organizations. Establishing and enforcing regulations that encourage sustainable practices is essential to protect both the space environment and Earth's atmosphere for future generations.

Conclusion:

Chapter 6 addresses the significant challenges and controversies associated with the commercialization of space travel. By examining safety concerns, regulatory challenges, ethical considerations, and the environmental impact of space activities, this chapter highlights the importance of responsible and sustainable practices in the private space industry. Addressing these challenges is crucial to ensuring the long-term success and acceptance of commercial space travel while safeguarding the interests of humanity and the environment.

CHAPTER 7:

ETHICAL AND LEGAL

CONSIDERATIONS

Exploitation of Space Resources:

As private companies and nations look to space for valuable resources, ethical considerations surrounding the exploitation of these resources become increasingly important. The potential for mining asteroids, the Moon, and other celestial bodies raises questions about sustainability and international cooperation.

Ethical Issues in Resource Extraction:

The extraction of space resources poses several ethical dilemmas. The potential for over-exploitation and environmental degradation must be carefully managed. Just as on Earth, responsible stewardship is crucial to ensure that space resources are used sustainably and equitably.

Sustainable Practices and Resource Management:

Implementing sustainable practices in space resource extraction is essential to prevent the depletion of these resources and protect the space environment. Companies must develop technologies and methods that minimize environmental impact and promote long-term sustainability.

International Agreements and Guidelines:

International cooperation is vital in managing space resources. The Outer Space Treaty of 1967, along with other agreements, provides a framework for the peaceful use of space. However, new guidelines and agreements may be needed to address the unique challenges posed by space resource extraction.

Colonization and Human Rights:

The prospect of colonizing other planets, particularly Mars, presents significant ethical and legal challenges. Ensuring that space colonies are governed by fair and just systems that protect human rights is crucial.

Ethical Implications of Space Colonization:

Space colonization raises questions about the ethical implications of human settlement on other planets. Issues such as the impact on potential extraterrestrial life and the preservation of pristine environments must be considered.

Ensuring Human Rights in Space Colonies:

Establishing colonies on other planets necessitates the development of legal and governance systems that protect the rights of settlers. Ensuring access to basic necessities, freedom, and fair treatment is essential for the success and ethical integrity of space colonies.

Governance and Legal Systems for Space Settlements:

Developing effective governance structures for space settlements is a complex task. Legal systems must be designed to handle disputes, ensure justice, and promote peaceful coexistence among settlers. International collaboration will be key to creating these frameworks.

Military Use of Space:

The potential for the militarization of space poses significant ethical and legal concerns. Preventing the deployment of weapons in space and ensuring that space remains a realm of peaceful exploration is a priority for the global community.

Potential Militarization of Space:

The use of space for military purposes, including the deployment of weapons, poses a threat to international peace and security. The potential for conflicts in space necessitates strict regulations and preventive measures.

Ethical Concerns and Preventive Measures:

Ethical concerns about the militarization of space include the risk of escalating conflicts and the potential for widespread destruction. Preventive measures, such as international treaties and agreements, are essential to maintaining the peaceful use of space.

International Treaties and Agreements:

Treaties like the Outer Space Treaty and the Anti-Ballistic Missile Treaty play a crucial role in preventing the militarization of space. Ongoing international dialogue and cooperation are necessary to strengthen these agreements and address emerging threats.

Legal Frameworks Governing Private Space Activities:

The rise of private space companies presents new challenges for existing legal frameworks. Adapting and developing space laws to regulate private activities and ensure compliance with international standards is essential.

Existing Space Laws and Regulations:

Current space laws, such as the Outer Space Treaty, provide a foundation for regulating space activities. However, these laws were primarily designed for state actors and may not fully address the complexities of private space ventures.

Challenges in Regulating Private Space Activities:

Regulating private space activities involves addressing issues such as liability, property rights, and the commercialization of space. Ensuring that private companies adhere to international standards and operate responsibly is a significant challenge for lawmakers.

Future Directions for Space Law:

As the space industry evolves, so too must space law. Developing new regulations that address the unique challenges of private space activities, promote sustainability, and ensure the peaceful use of space will be crucial for the future of space exploration.

CHAPTER 8:

VISION FOR THE FUTURE

Vision for the Next Decade:

The next decade promises significant advancements in space exploration, driven by both public and private sectors. Companies like SpaceX, Blue Origin, and new entrants aim to achieve milestones that were once considered science fiction.

Upcoming Milestones and Projects:

Significant projects on the horizon include SpaceX's Mars missions, Blue Origin's lunar lander, and NASA's Artemis program aiming to return humans to the Moon. These missions will lay the groundwork for more ambitious endeavors in the future.

Technological Advancements and Innovations:

Advances in propulsion systems, artificial intelligence, and sustainable technologies will drive the next wave of space exploration. These innovations will make space travel more efficient, safer, and more accessible.

Expanding the Boundaries of Human Exploration:

As we push the boundaries of human exploration, missions to Mars, asteroids, and beyond will become more feasible. These missions will not only expand our understanding of the universe but also pave the way for future generations of explorers.

Potential for Commercial Space Stations and Lunar Bases:

Commercial space stations and lunar bases represent the next frontier for space exploration and industry. These developments will support scientific research, tourism, and commercial activities in space.

Development of Commercial Space Stations:

Companies like Axiom Space and Bigelow Aerospace are developing commercial space stations that will provide habitats for researchers, tourists, and industrial activities. These stations will enable long-duration missions and support a wide range of scientific experiments.

Establishing Lunar Bases for Research and Industry:

Lunar bases will serve as hubs for scientific research and industrial activities, including resource extraction and manufacturing. Collaboration between public and private sectors will be key to developing the infrastructure needed for sustainable lunar operations.

Collaboration Between Public and Private Sectors:

Successful development of space stations and lunar bases will require close collaboration between government agencies and private companies. Public-private partnerships will facilitate the sharing of resources, expertise, and technology.

Long-Term Goals and Aspirations:

The long-term vision for space exploration includes human settlement on Mars, interplanetary travel, and the establishment of permanent colonies beyond Earth.

Human Settlement on Mars and Beyond:

Human settlement on Mars is a primary goal for many space exploration initiatives. Establishing self-sustaining colonies on Mars will be a monumental achievement, requiring advancements in life support, habitat construction, and resource utilization.

Interplanetary Travel and Exploration:

Future missions may extend beyond Mars, exploring the outer planets and their moons. Interplanetary travel will open new frontiers for scientific discovery and human expansion.

The Role of International Collaboration:

International collaboration will be crucial for the success of these long-term goals. Joint missions, shared research, and coordinated efforts will help humanity achieve its aspirations in space.

Inspiring the Next Generation:

Inspiring the next generation of scientists, engineers, and explorers is essential for the continued success of space exploration. Educational programs, mentorship opportunities, and a culture of innovation will foster the development of future space pioneers.

Educational Programs and Outreach:

Educational programs that highlight the achievements and possibilities of space exploration will inspire young minds. Outreach initiatives can engage students and provide them with the knowledge and skills needed for careers in space.

Mentorship and Scholarship Opportunities:

Mentorship programs that connect students with professionals in the space industry can provide valuable guidance and support. Scholarships can help students pursue advanced degrees in STEM fields, ensuring a steady pipeline of talent for the space industry.

Fostering a Culture of Innovation and Exploration:

Encouraging innovation and a spirit of exploration will be key to advancing space technology and achieving future milestones. Competitions, grants, and other incentives can promote creative solutions to the challenges of space exploration.

Conclusion:

Chapter 8 looks forward to the future of space exploration, outlining the vision for the next decade, the potential for commercial space stations and lunar bases, long-term goals, and the importance of inspiring the next generation. By embracing innovation, fostering collaboration, and nurturing the next generation of space enthusiasts, humanity can continue to push the boundaries of what is possible in the final frontier.

CONCLUSION

As we stand on the brink of a new era in space exploration, the achievements and aspirations of private space companies herald a transformative period in human history. The journey from government-led missions to the dynamic involvement of private enterprises has opened up unprecedented opportunities for scientific discovery, technological innovation, and the expansion of human presence beyond Earth.

Our exploration began with the dawn of private spaceflight, where visionary entrepreneurs like Elon Musk, Jeff Bezos, and Richard Branson dared to dream of a future where space is accessible to all. Their pioneering efforts laid the foundation for an industry that is now thriving and pushing the boundaries of what was once thought possible. From SpaceX's historic Falcon 1 launch to Blue Origin's New Shepard milestones and Virgin Galactic's strides in space tourism, these missions have not only demonstrated technological prowess but also ignited the public's imagination.

Technological innovations have been at the heart of this transformation. Reusable rockets, advanced propulsion systems, and innovations in life support and space habitats are making space travel more efficient, sustainable, and within reach. These advancements are driving the space industry forward, enabling missions that were once the stuff of science fiction. The economic implications are profound, with substantial investments and cost reduction strategies positioning the private space sector as a significant driver of global economic growth. The impact of space tourism on public interest and industry growth underscores the transformative power of human spaceflight, showcasing the personal experiences of private astronauts and space tourists.

However, this journey is not without its challenges. Safety concerns, regulatory hurdles, ethical considerations, and environmental impacts are critical issues that must be addressed to ensure the sustainable and ethical exploration of space. The potential exploitation of space resources, the militarization of space, and the governance of human rights in space colonies present complex dilemmas that require responsible practices and international cooperation. The evolving legal frameworks governing private space activities are crucial for maintaining the peaceful and equitable use of outer space.

Looking to the future, the vision for the next decade includes the development of commercial space stations, lunar bases, and the human settlement on Mars and beyond. The potential for interplanetary travel and the establishment of permanent colonies will open new frontiers for scientific discovery and human expansion. International collaboration, educational programs, and mentorship opportunities will play a pivotal role in inspiring the next generation of scientists, engineers, and explorers.

Our previous book, "MARS Colonization: The Red Planet New Era," laid the groundwork for understanding the possibilities and challenges of establishing a human presence on Mars. This current exploration builds upon that foundation, expanding our horizons to encompass the broader landscape of private spaceflight and the myriad opportunities it presents.

As we look to the future, the collaboration between public and private sectors, the relentless pursuit of innovation, and the enduring human spirit of exploration will drive us forward. Together, we will continue to explore, discover, and thrive in the vast expanse of space, pushing the boundaries of what is possible and ensuring that the final frontier remains a realm of opportunity and inspiration for all of humanity. The journey ahead is filled with challenges, but it is also brimming with promise. By learning from our past, embracing the present, and envisioning a future where space is within everyone's reach, we can achieve remarkable things. The stars are no longer out of reach; they are the next destination on our journey of exploration and discovery.

The milestones we achieve in space exploration today will shape the future of humanity in profound ways. Every successful mission, every technological breakthrough, and every step towards sustainability brings us closer to a future where space is an integral part of human existence. The advancements made by private space companies are not just about reaching new heights; they are about creating new possibilities and opportunities for generations to come.

The importance of inspiring the next generation cannot be overstated. As we venture further into space, it is imperative that we engage young minds and nurture their curiosity and passion for exploration. Educational programs, outreach initiatives, and mentorship will be key to fostering the next wave of scientists, engineers, and innovators who will carry the torch of space exploration forward. These young pioneers will be the ones to build upon the foundation we lay today, pushing the boundaries even further and unlocking the mysteries of the cosmos.

International collaboration will also play a critical role in the future of space exploration. By working together, nations can pool their resources, expertise, and knowledge to achieve greater things than any single country could accomplish alone. Joint missions, shared research, and coordinated efforts will not only advance our understanding of space but also promote peace and cooperation on Earth. The collaborative spirit that drives space exploration can serve as a model for addressing global challenges and fostering unity among nations.

The long-term goals of space exploration extend beyond technological advancements and economic growth. They encompass the broader vision of ensuring the survival and prosperity of humanity. As we establish permanent colonies on other planets and explore the potential for interplanetary travel, we are taking steps towards securing a future for our species beyond the confines of Earth. These endeavors will require innovative solutions to the challenges of living in space, from life support systems and habitat construction to sustainable resource management and governance.

The ethical considerations of space exploration are equally important. As we venture into new frontiers, we must do so with a commitment to preserving the integrity of extraterrestrial environments and ensuring that our actions do not cause harm. The responsible and equitable use of space resources, the protection of human rights in space colonies, and the prevention of militarization are all critical issues that must be addressed through international agreements and ethical guidelines.

In conclusion, the journey of space exploration is one of the most exciting and significant endeavors of our time. The progress made by private space companies, the collaborative efforts of international partners, and the inspiration drawn from these achievements are shaping a future where space is an integral part of human existence. As we look to the stars, we are reminded of the boundless potential of human ingenuity and the limitless possibilities that lie ahead. The final frontier is not just a destination; it is a symbol of our collective aspiration to explore, discover, and thrive in the universe. By embracing this vision, we can ensure that the legacy of space exploration continues to inspire and uplift humanity for generations to come.

FURTHER READING AND RESOURCES

For readers who wish to delve deeper into the fascinating world of space exploration, private spaceflight, and the future of humanity beyond Earth, the following resources provide a wealth of information and insights:

Books:
1- "The Space Barons: Elon Musk, Jeff Bezos, and the Quest to Colonize the Cosmos" by Christian Davenport (PublicAffairs)

2- "Rocket Men: The Daring Odyssey of Apollo 8 and the Astronauts Who Made Man's First Journey to the Moon" by Robert Kurson (Random House)

3- "How to Build a Spaceship: A Band of Renegades, an Epic Race, and the Birth of Private Spaceflight" by Julian Guthrie (Penguin Press)

4- "MARS Colonization: The Red Planet New Era" by Rayan Bale

5- "Endurance: A Year in Space, A Lifetime of Discovery" by Scott Kelly (Knopf)

Websites and Online Resources:

1- NASA (National Aeronautics and Space Administration)
   o www.nasa.gov

2- SpaceX
   o www.spacex.com

3- Blue Origin
   o www.blueorigin.com

4- Virgin Galactic
   o www.virgingalactic.com

5- The Planetary Society
   o www.planetary.org

Documentaries and Films:

1- "Apollo 11" (2019)

2- "The Mars Generation" (2017)

3- "For All Mankind" (1989)